What if?

Keith Pruitt, Ed. S

Words of Wisdom
2019

What if there was a rainbow in
the sky all the time?

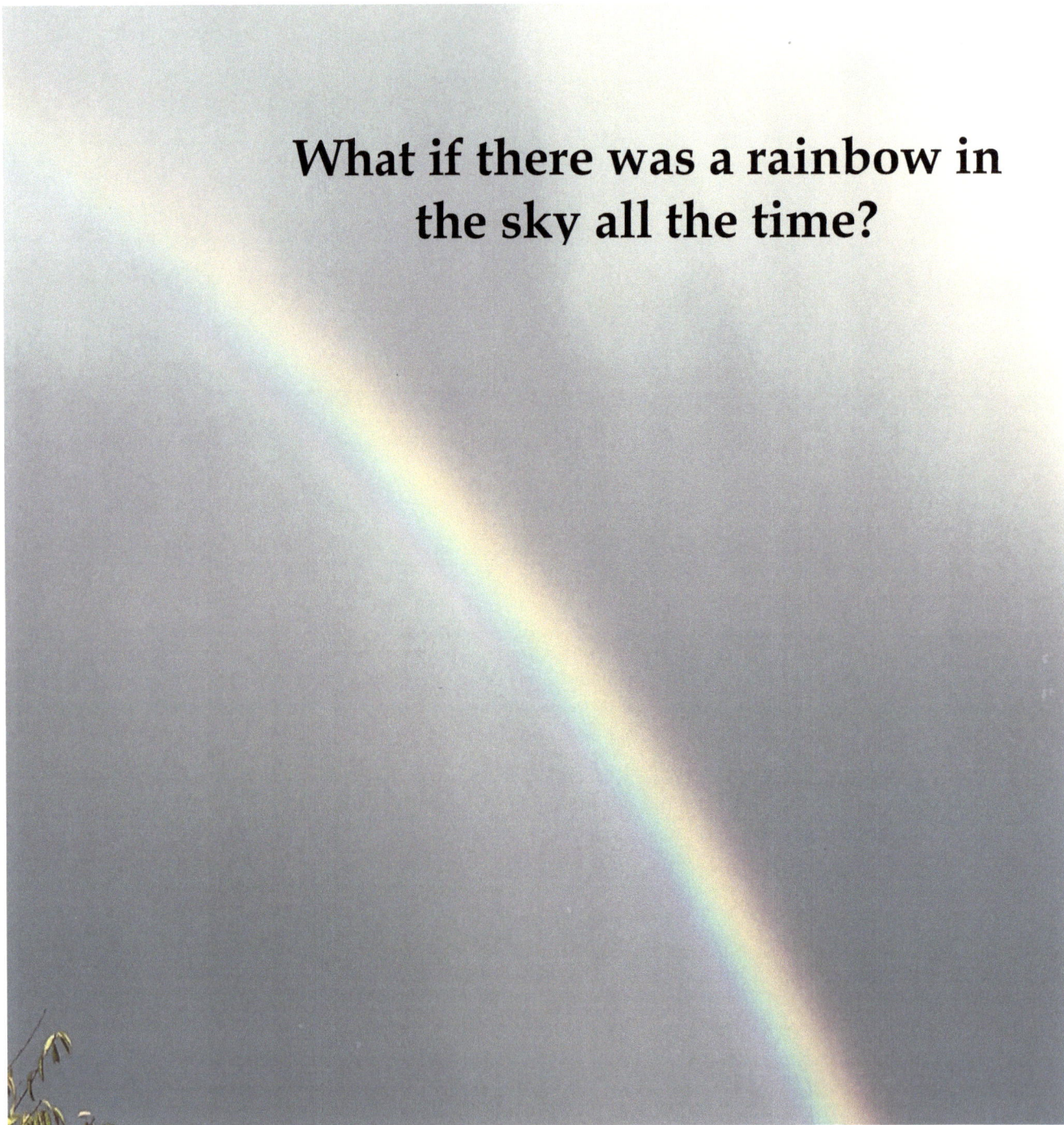

What if the whole world were covered in water?
Where would we live?

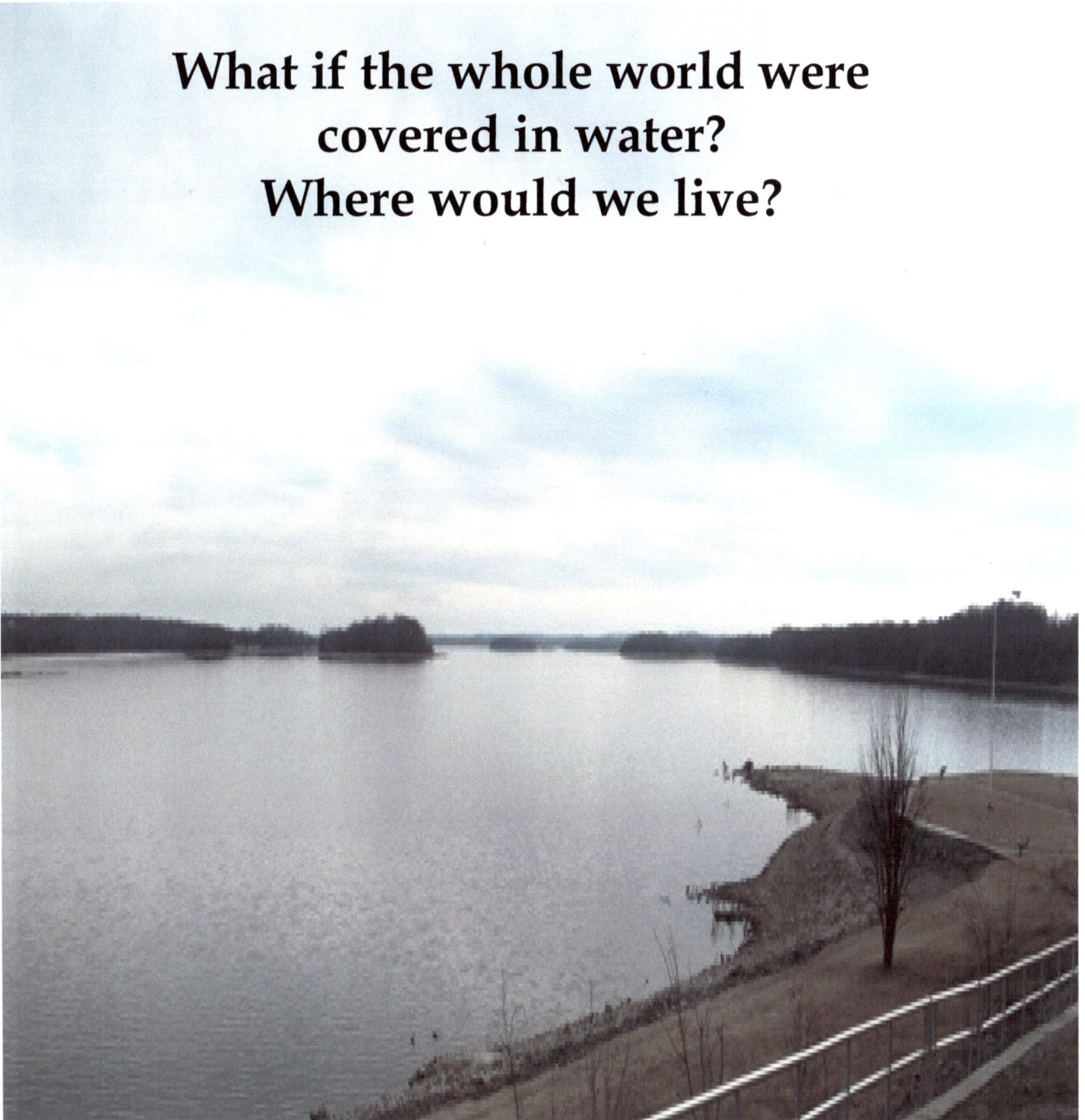

What if instead of talking we all quacked like ducks?

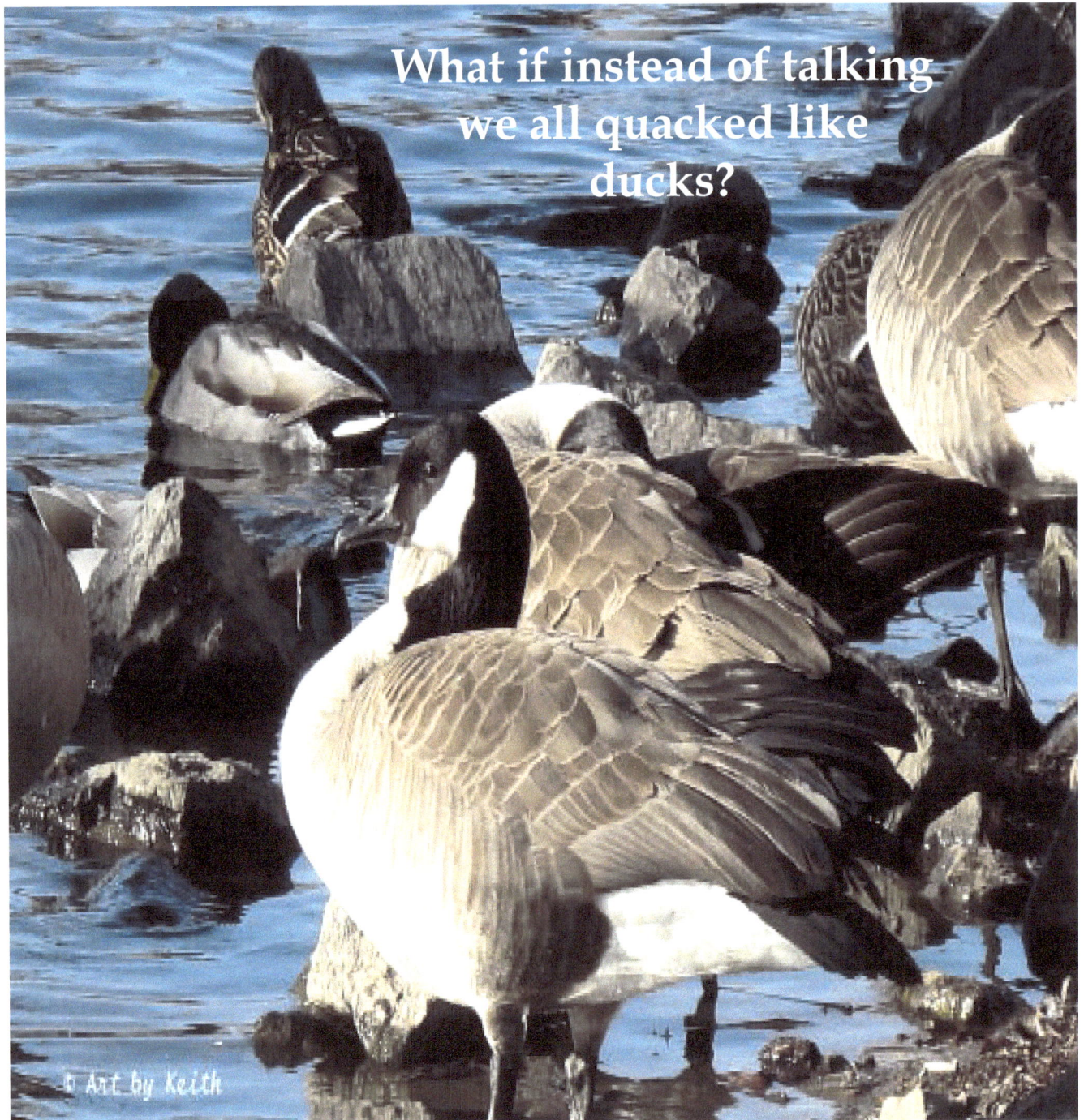

© Art by Keith

What if we all had wings like eagles?

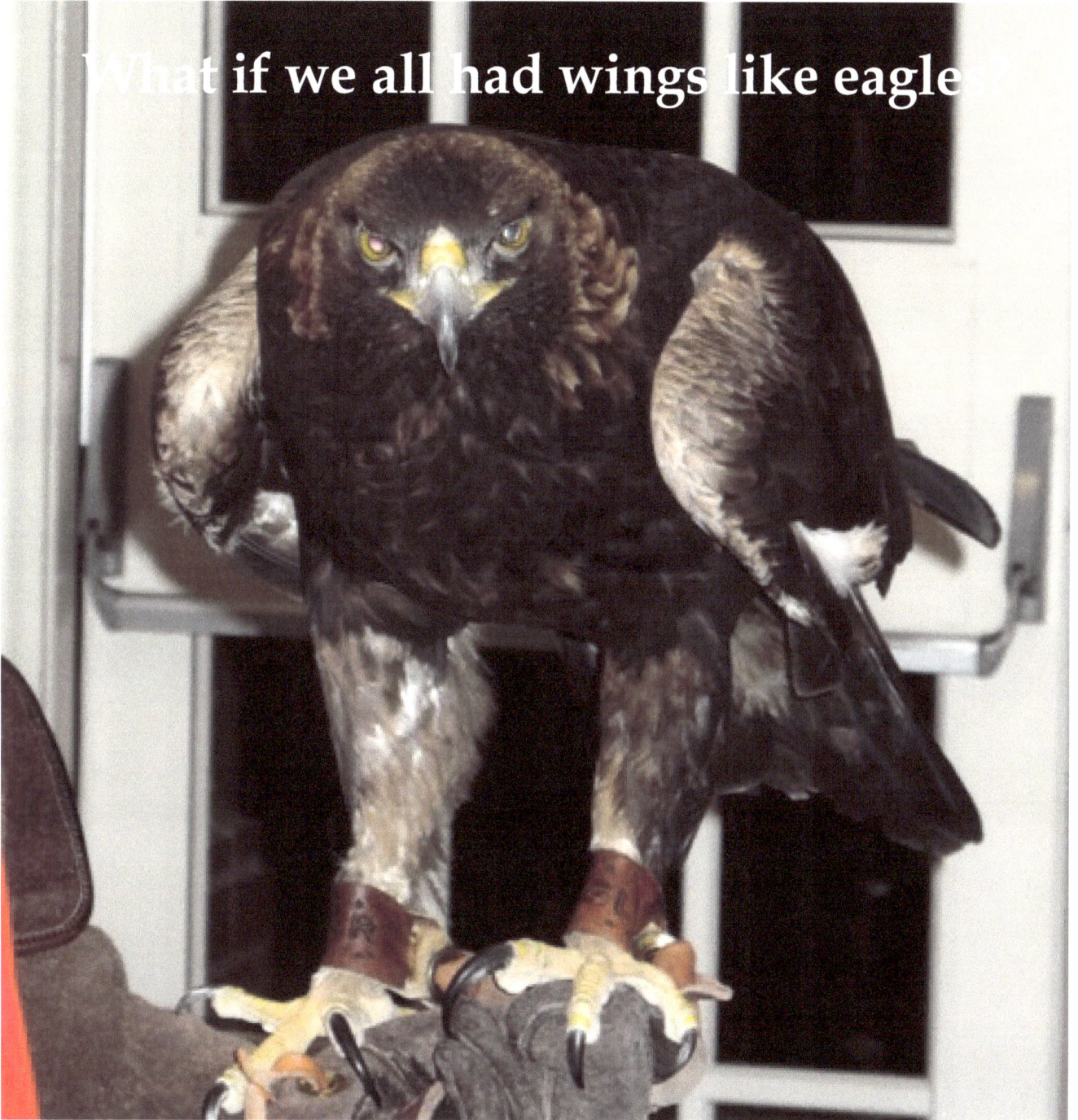

What if trees grew upside down?

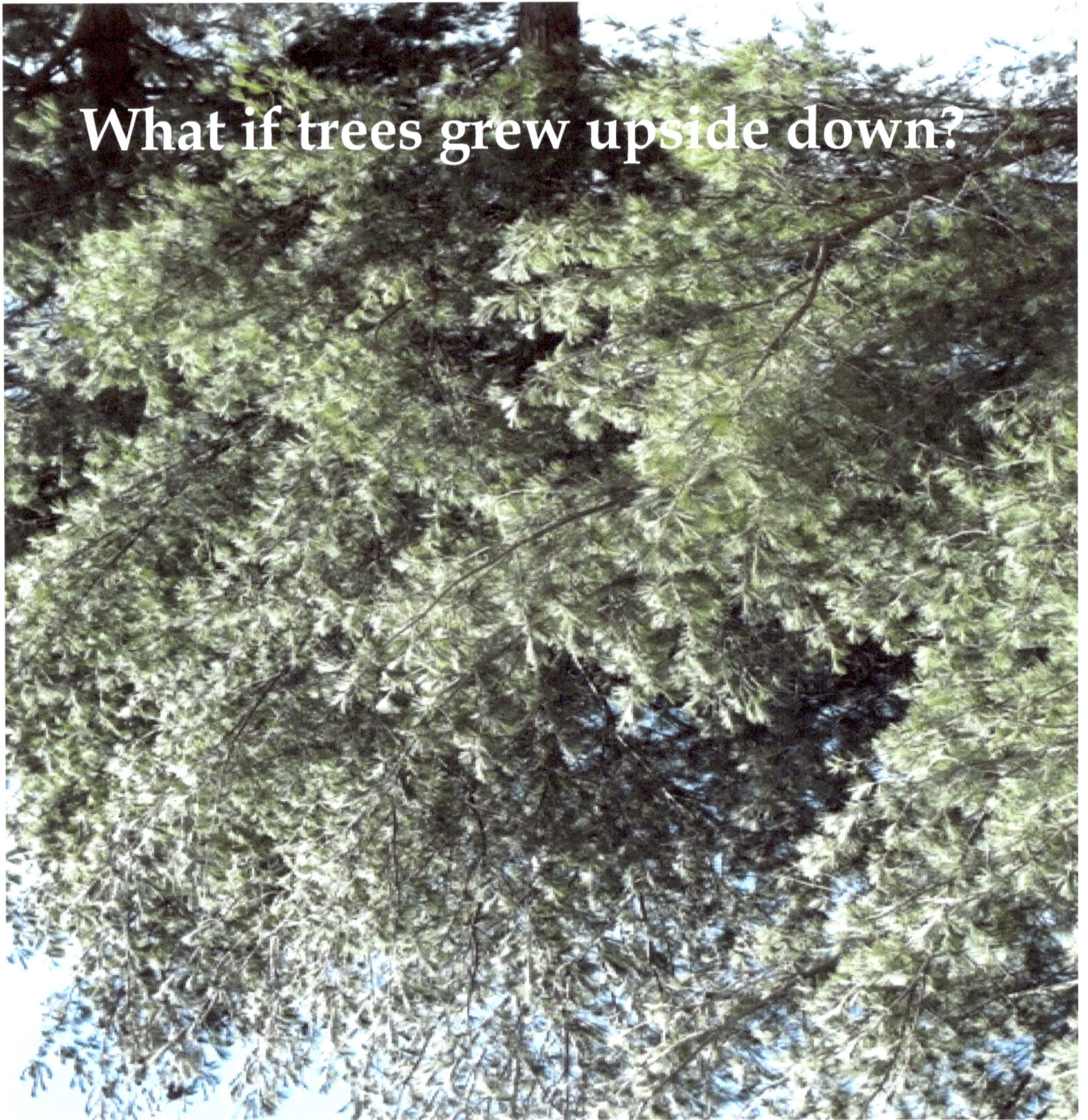

What if there was no more oil?

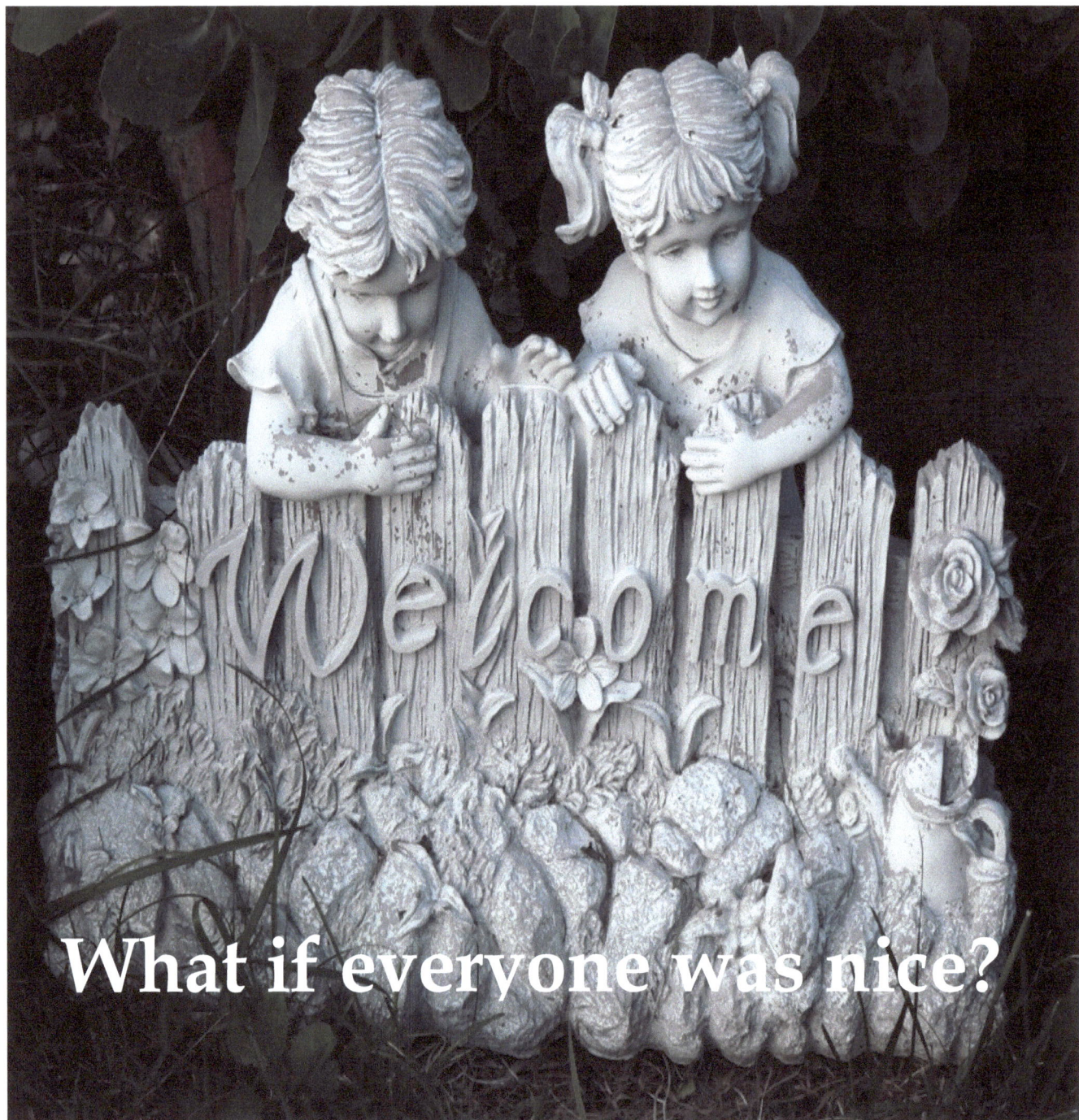

What if everyone was nice?

What If There Was No Color?

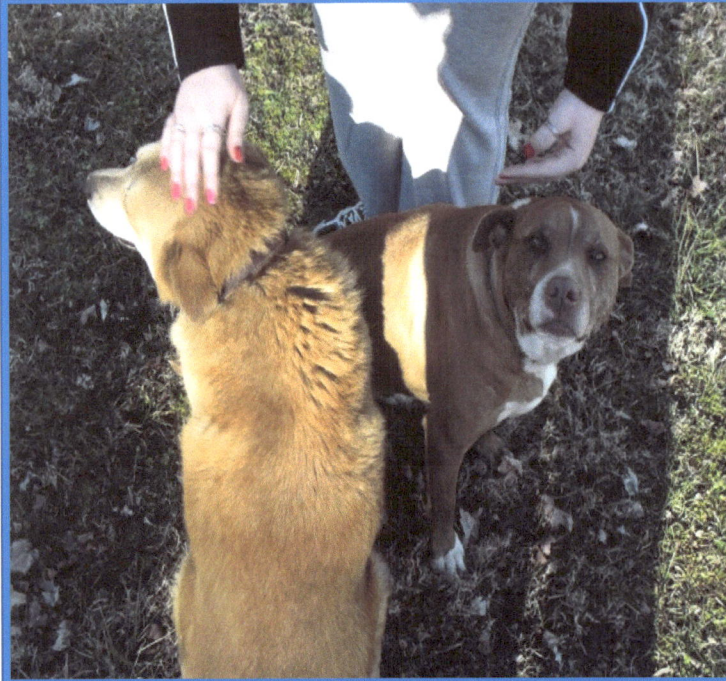

What if dogs meowed?

And Cats Barked!

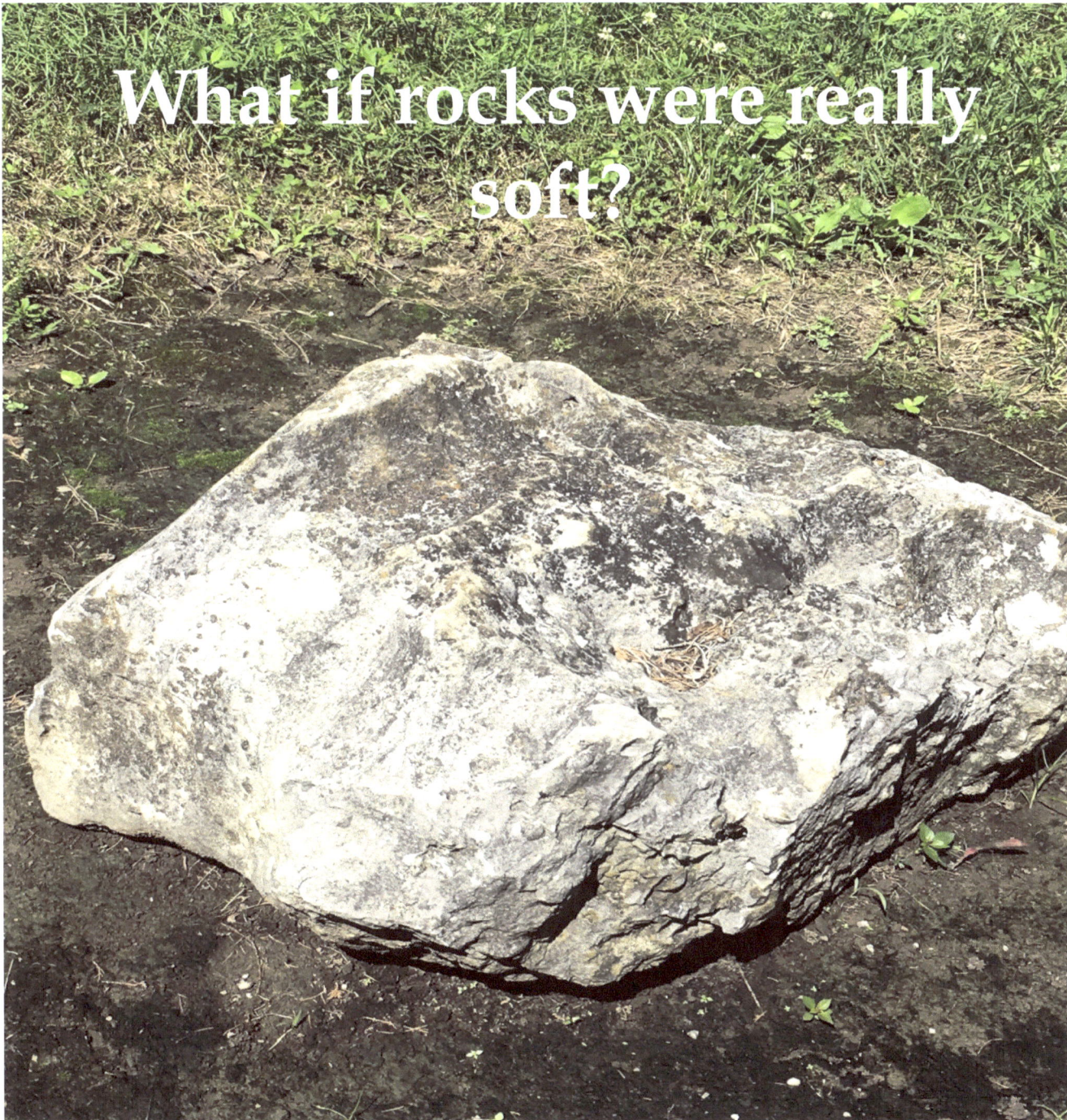

What if rocks were really soft?

What if Tigers were really tame?

What if sea turtles could fly?

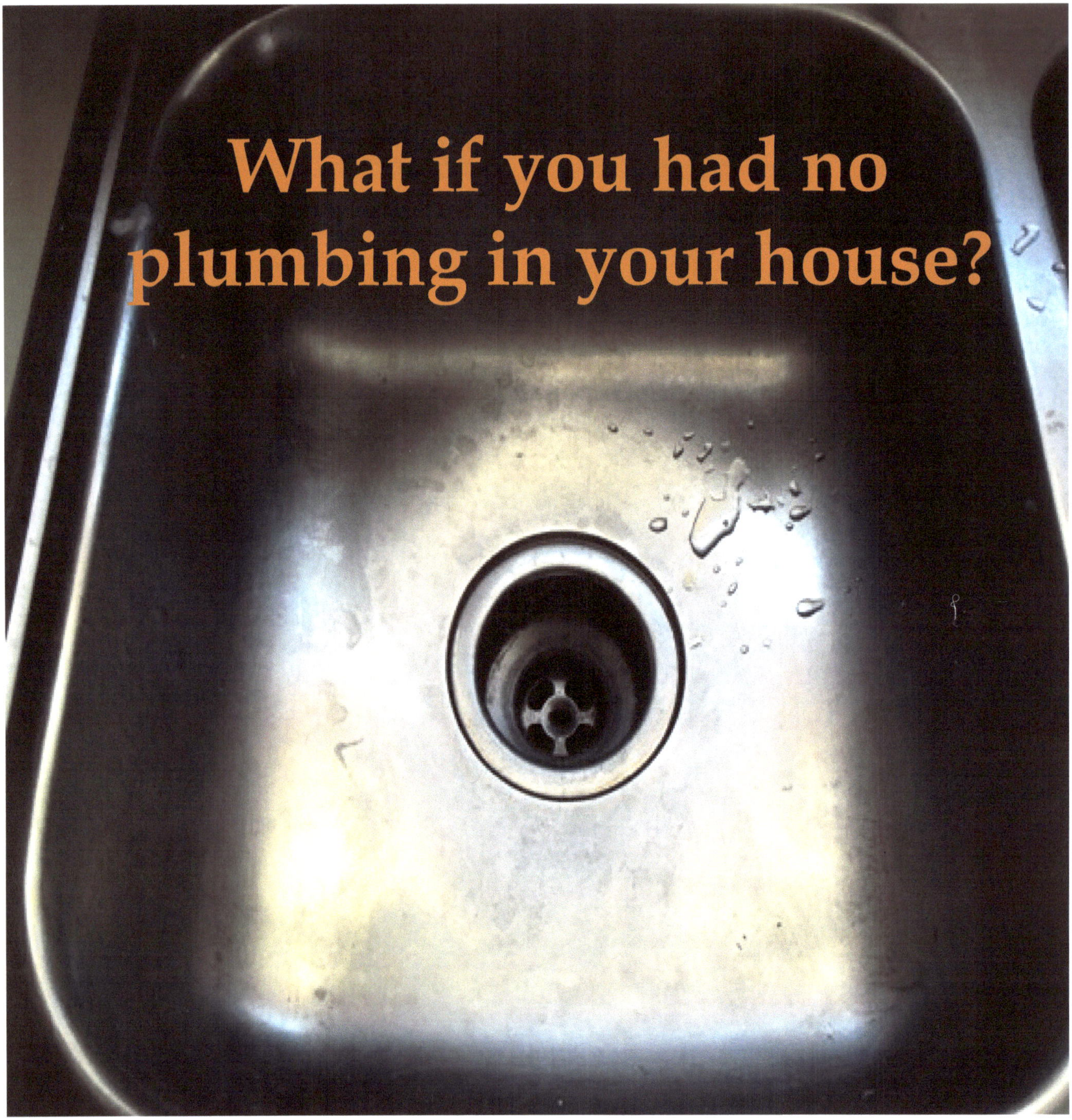
What if you had no
plumbing in your house?

What if monkeys lived in houses, and people lived in trees?

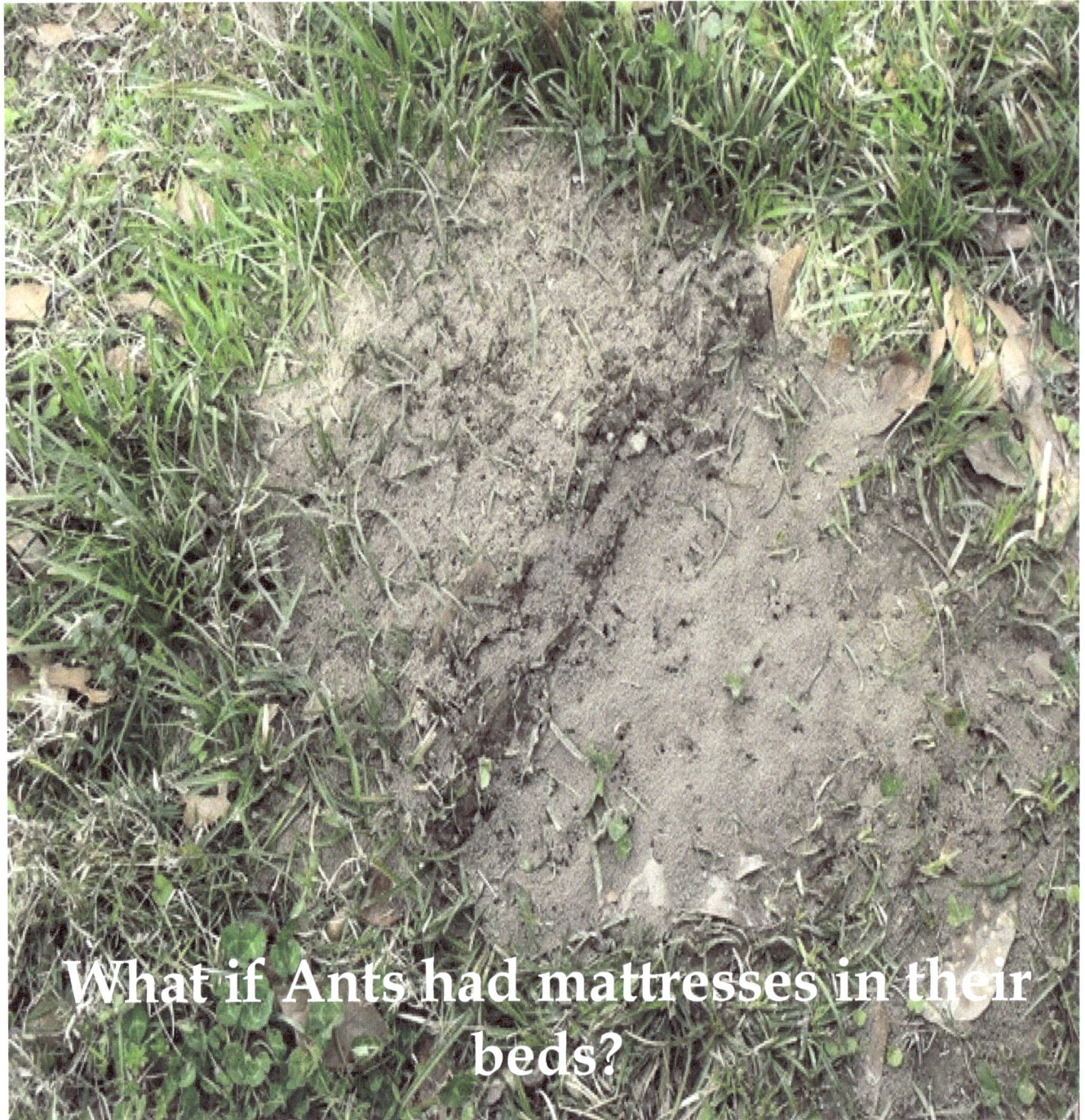

What if Ants had mattresses in their beds?

What if fog were on a mountain

In Winter?

What if you lived at the ocean?

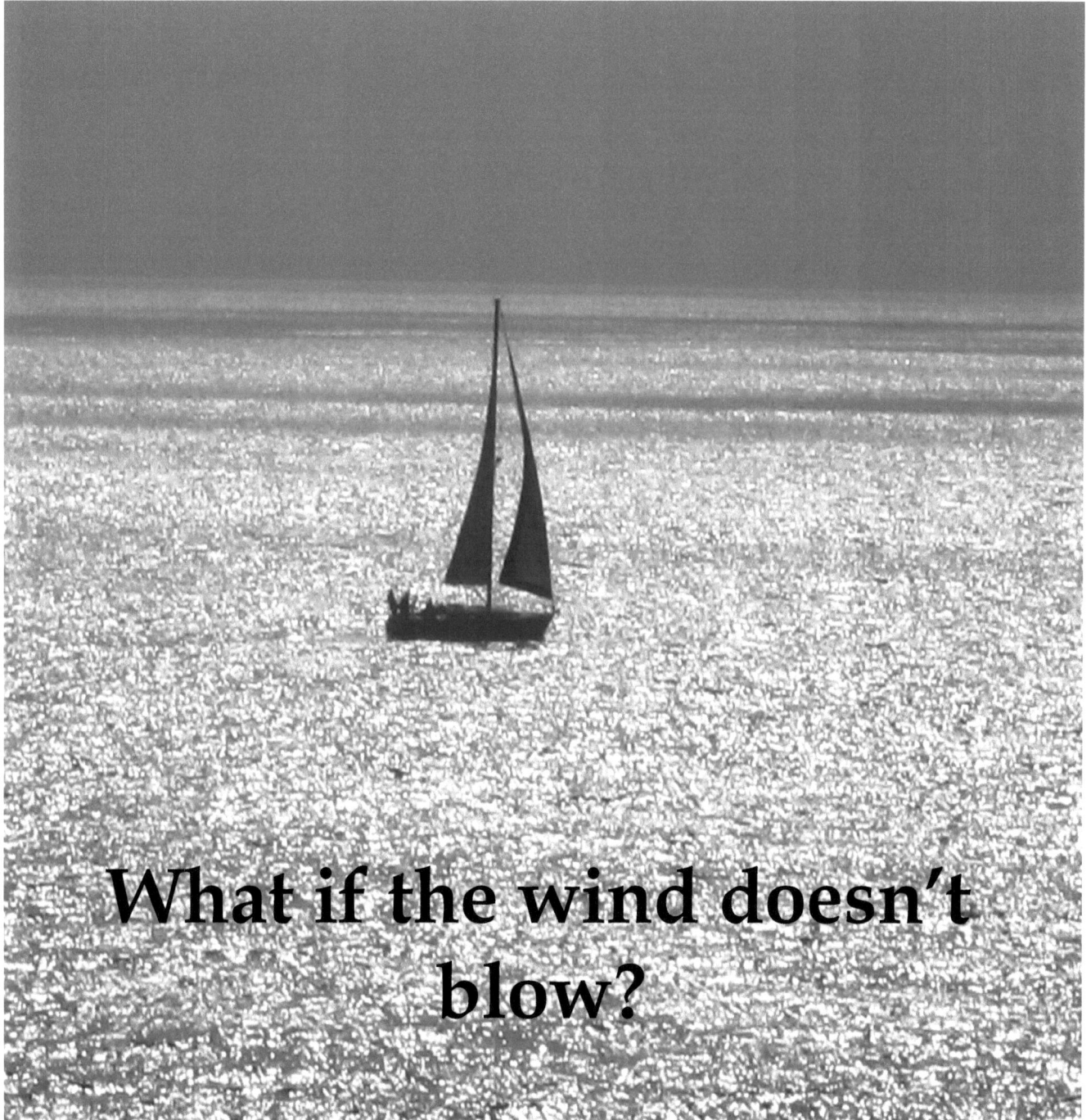

What if the wind doesn't blow?

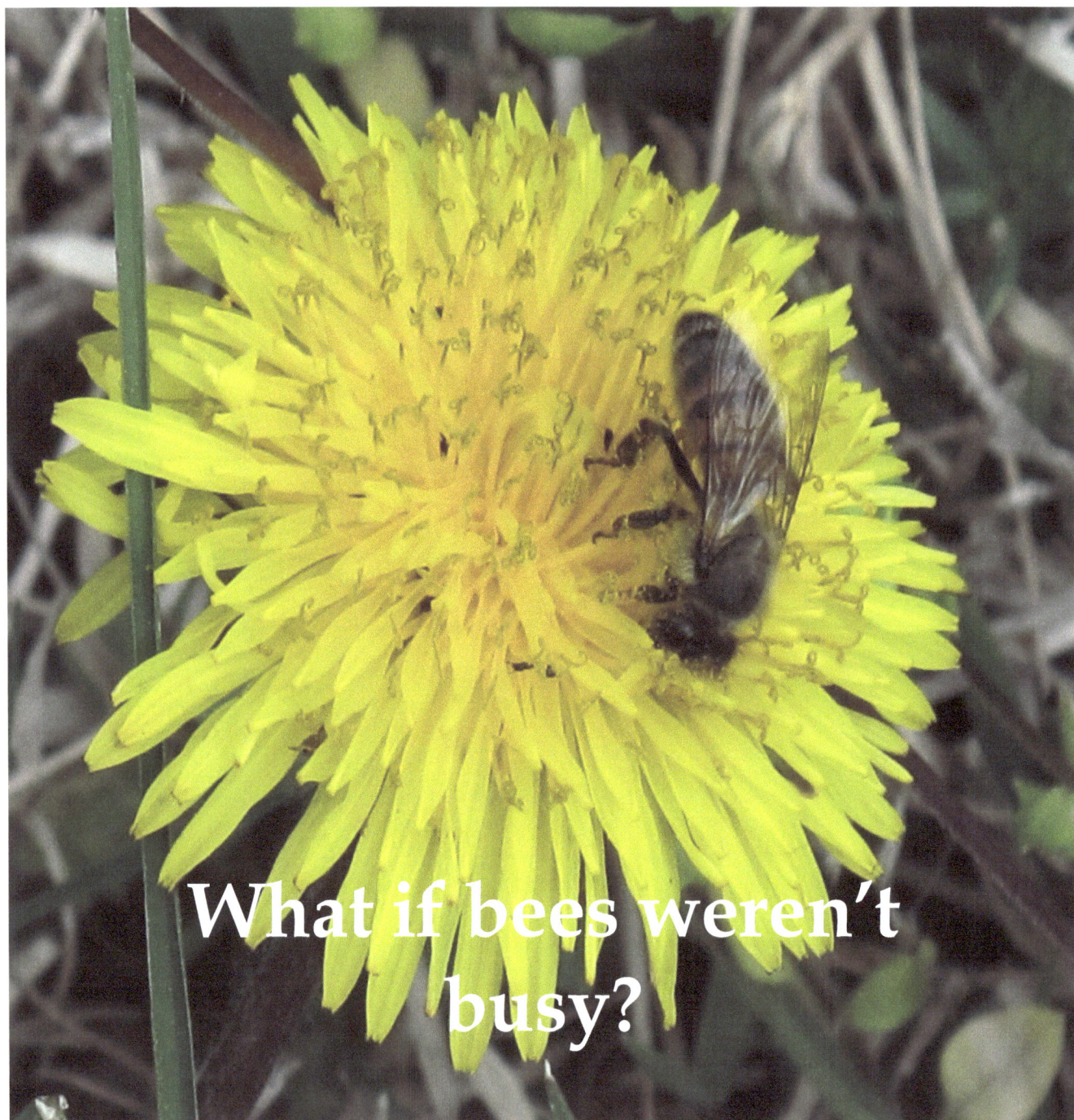

What if bees weren't busy?

What if you could swim like fish?

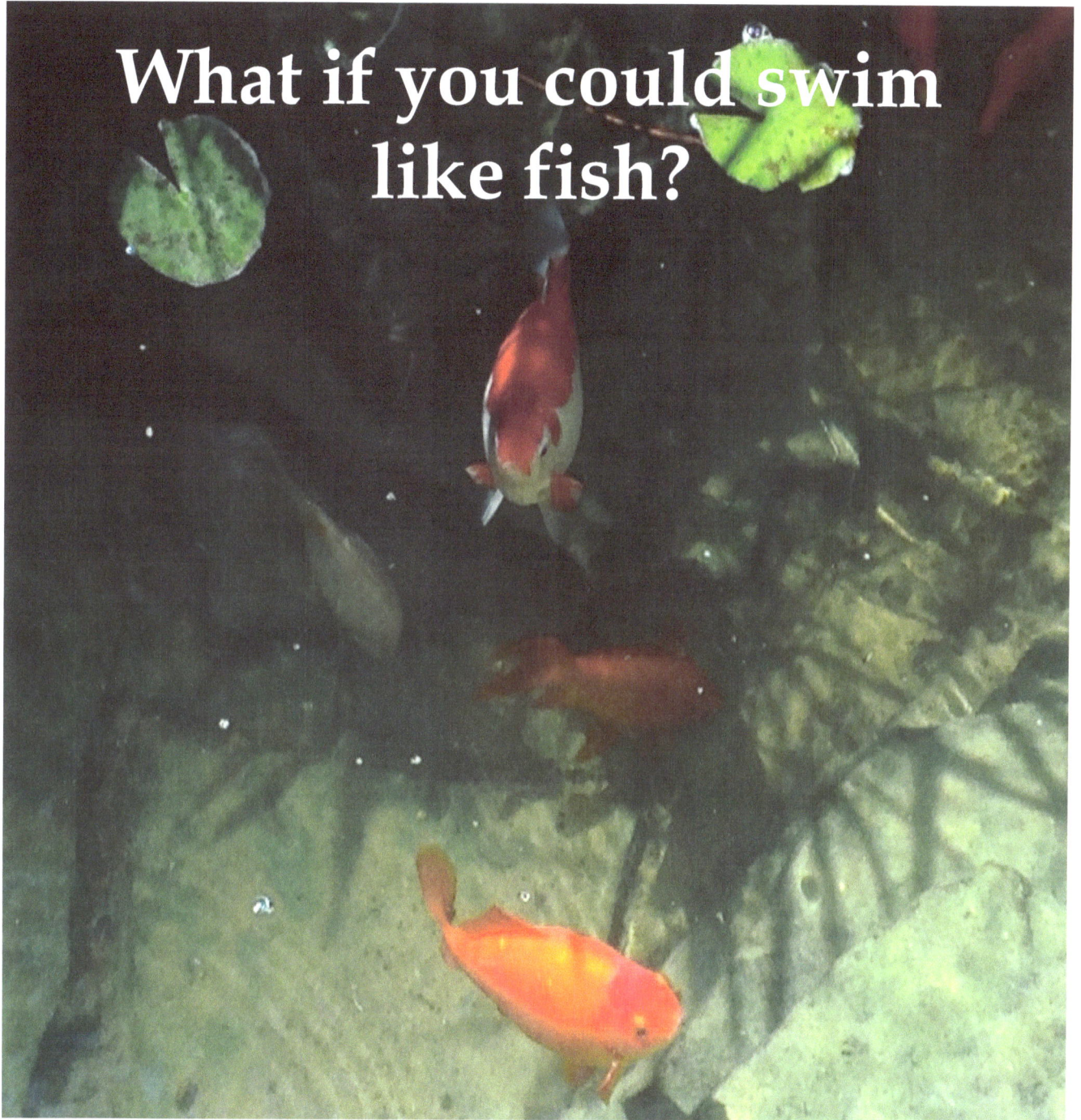

What if leaves floated in the air instead of falling to the ground?

What if all streets were one way?

What if you had as many legs as a spider?

Garden Spider

What if you walked as many miles as a butterfly can fly?

What if you hopped like a rabbit?

What if you were as tall as a giraffe?

What if you bloomed like this cactus?

A Note from the Author

Thinking about what is possible or even the impossible can help us understand the world around us. In this whimsical look at our world, it is hoped you've learned a bit about the world around us. From learning about gravity to plant behavior to human characteristics, the author has brought to your attention the world we live in.

We always need to be thoughtful of this vast planet in which we live and the unique characteristics of even the tiniest of life on it. See if you can add to these what ifs... What if Ice Cream didn't melt. What if show was as tall as me? What if it didn't rain for like five days? What if I could eat flowers? What if... well, you get the idea. So, remember, I can learn a lot by just asking What If?

Keith Pruitt

July 3, 2019

www.ingramcontent.com/pod-product-compliance
Lightning Source LLC
Chambersburg PA
CBHW052046190326
41520CB00003BA/209